POSTCARDS from the LAKE

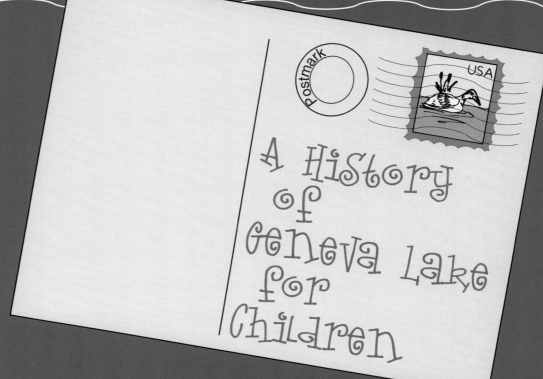

A History
of
Geneva Lake
for
Children

Postmark

USA

Cynthia Kelly Conlon, J.D., Ph.D.

This book is dedicated to
my husband, Bob,
and three children,
Kevin, Caitlin, and Molly,
who love the lake.

Postmark

LOVE
USA

Copyright 2001 by Cynthia Kelly Conlon.

Library of Congress Control Number 2001091777
ISBN 0-9711736-0-5

Published by Geneva Lake Publishing, Lake Geneva, Wisconsin.
Printed in the United States of America.

Publication of this book has been made possible,
in part, through the support of the
Geneva Lake Conservancy.

Historic postcards were reproduced through the generosity
of the Geneva Lake Area Museum of History.

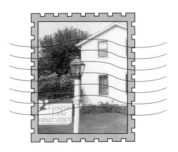

Author and Student Illustrators from St. Francis de Sales Parish School, Lake Geneva, Wisconsin.

Amanda Bergesen, 5th grade, page 8; Bobby Callan, 2nd grade, page 20; Jimmy Callan, kindergarten, page 15; Caitlin Conlon, 5th grade, cover, and pages, 1,4, and 10; Kevin Conlon, 6th grade, inside back cover and page 23; Molly Conlon, 4th grade, pages 2, 9, and 17; Hannah Green, 3rd grade, page 6; Kolin Harrigan, 2nd grade, page 11; Sarah Harrigan, 5th grade, page 16; Alex Heinz, 4th grade, page 18; Frances Homan, 2nd grade, page 25; Ian Johnson, 2nd grade, page 14; Robert Johnson, 4th grade, page 21; Matty Kelly, 5th grade, page 26; Peter Leedle, 5th grade, page 19; Andy Mackey, 5th grade, page 27; Andrea Miller, 6th grade, page 12; Lee Pankau, 4th grade, page 22; Chris Shields, 6th grade, page 13; Ryan Thiel, 5th grade, page 24; Charlie Williams, 6th grade, page 7; Kevin Williams, 3rd grade, page 5.

~Post Card~

Postmark

This is a story about a lake.

You may think that it is odd to read a story in which a lake is the main character, but stories about places are important. Stories about places help us understand who we are, where we came from, and what we are in the process of becoming.

The lake in this story is located in southeastern Wisconsin and is called Geneva Lake. You may have also heard it called Lake Geneva, but Lake Geneva is really the name of the city on the northeastern shore of the lake.

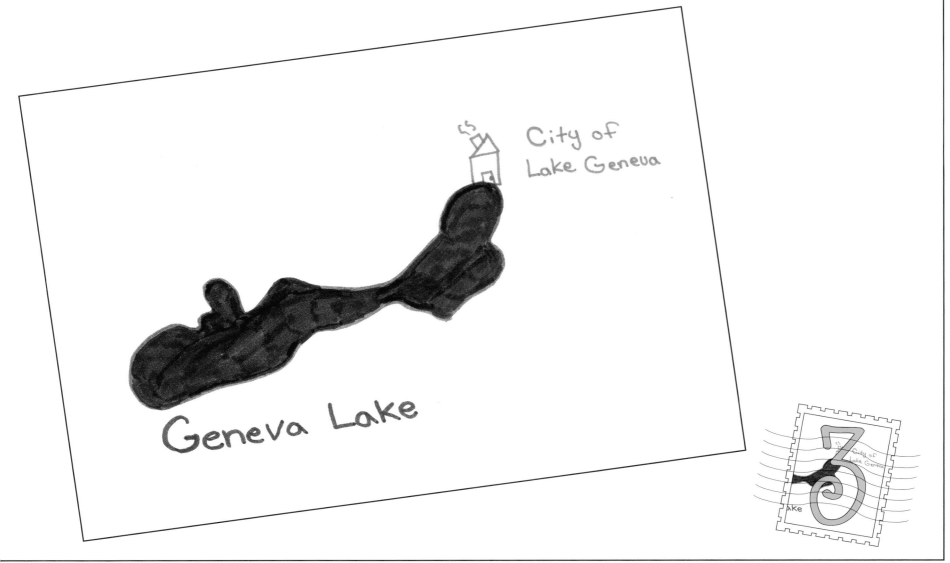

Geneva Lake is big, deep, and beautiful. It is so big that it would take you at least eight hours to walk all the way around it. It is so deep that you would need scuba diving equipment to reach the bottom!

Fun Facts:

The shore of the lake is 20.2 miles long.

The lake is 144 feet deep.

It is 7.6 miles long but only one-half mile wide at its narrowest point.

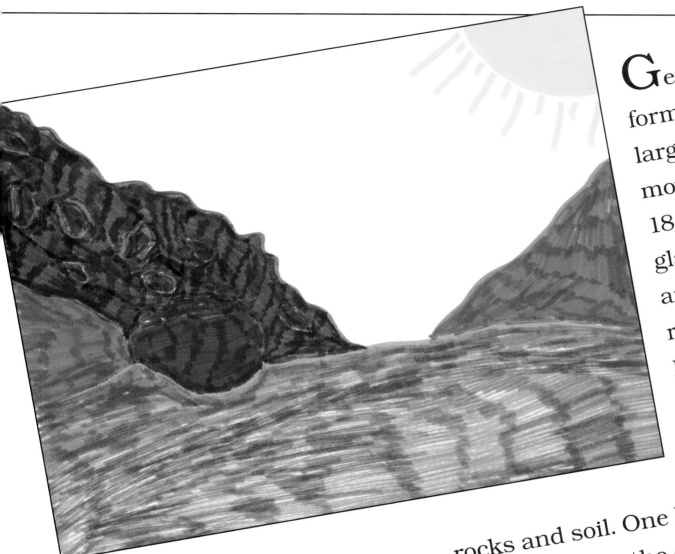

Geneva Lake was formed by glaciers, large fields of moving ice. About 18,000 years ago glaciers covered this area, scouring out river and lake beds. During periods of rising temperatures, ice melted and the glaciers deposited rocks and soil. One large chunk of ice that remained after the other ice had melted created the bed of Geneva Lake.

5

Geneva Lake's natural beauty has drawn people for hundreds of years. Some of the first people to live near the lake were Native Americans known as the Potawatomi tribe. The Potawatomis called the lake "Kishwauketoe" or "clear water." They lived in camps along the shore and created a walking path around the lake that is still used today.

~Post Card~

Fun Fact: Kishwauketoe Conservancy in Williams Bay uses this Potawatomi name.

One famous Potawatomi chief was named Big Foot. Legend says that he was given this name after he dragged a deer off the frozen lake one winter day and left behind a very large set of footprints.

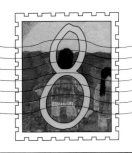

The Potawatomis considered Kishwauketoe their home, but they also believed that they shared the lake with the many plants, fish, and animals who also lived there.

In 1831, the first white people saw Geneva Lake. John Kinzie, his wife, Juliette, and some others were traveling from Fort Dearborn in Chicago to Fort Winnebago (at what today is Portage, Wisconsin). On the fourth day of this trip, they took a sudden turn and saw Geneva Lake. Juliette wrote about this moment in her diary, saying that a "shout of delight" burst from everyone's lips when the lake's "charming landscape" came into view.

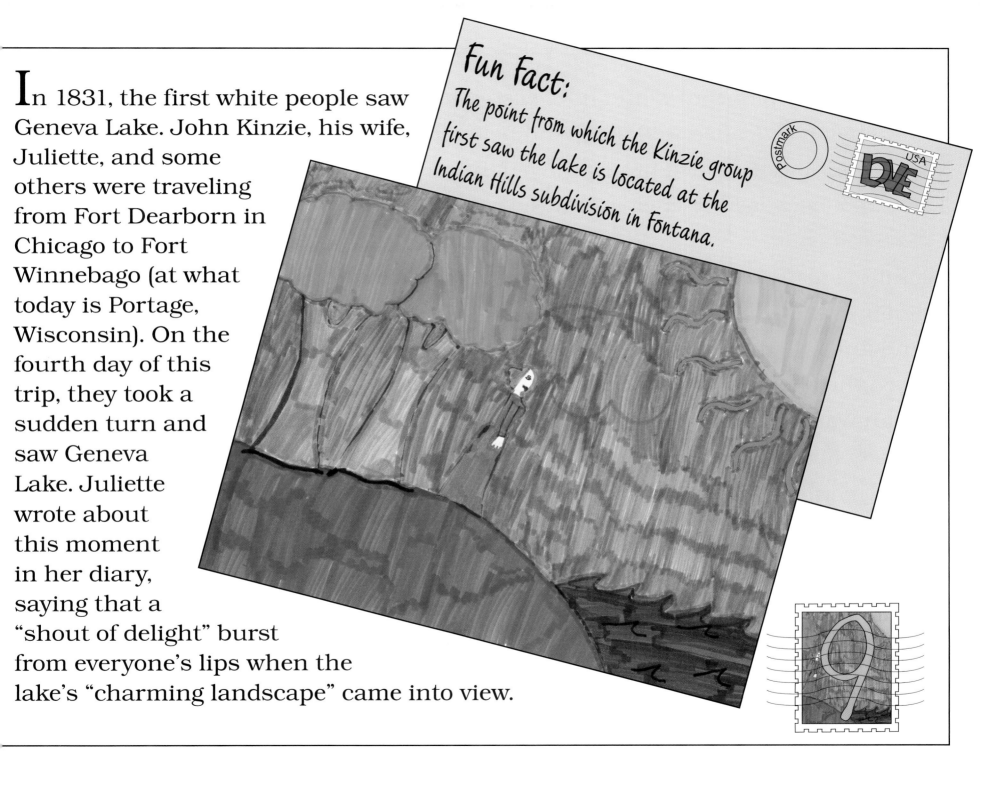

Fun Fact:
The point from which the Kinzie group first saw the lake is located at the Indian Hills subdivision in Fontana.

These early pioneers would have seen a lake surrounded by marshy areas (called wetlands) and fields filled with grasses and wildflowers (called prairies). Wetlands are important to the lake because they help clean the water that flows into it. Wetlands and prairies also are home to many plants and animals that need the shelter, clean water, and variety of foods that these communities provide.

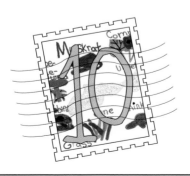

John Brink was one of the early pioneers who came to the Wisconsin territory. He was a surveyor who came in 1835 to make a map of the area for the government. Surveyors use special instruments to measure the area and elevation of land. He named the lake "Geneva" because it reminded him of a lake near his home town in Geneva, New York.

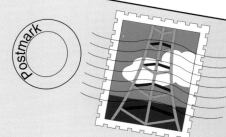

Under the terms of a government treaty, the Potawatomis were removed to a reservation in Kansas in 1836. Land around Geneva Lake was offered for sale at $1.25 an acre and many people bought land for farming. Other early pioneers logged the heavily wooded countryside.

In 1858 the era of steamboats on the lake began with the arrival of the Atalanta. The Atalanta was used to take sightseers along the shore. Soon many commercial and private steamers were in use. Later, larger, double-decker steamboats such as the Lady of the Lake held even larger numbers of passengers.

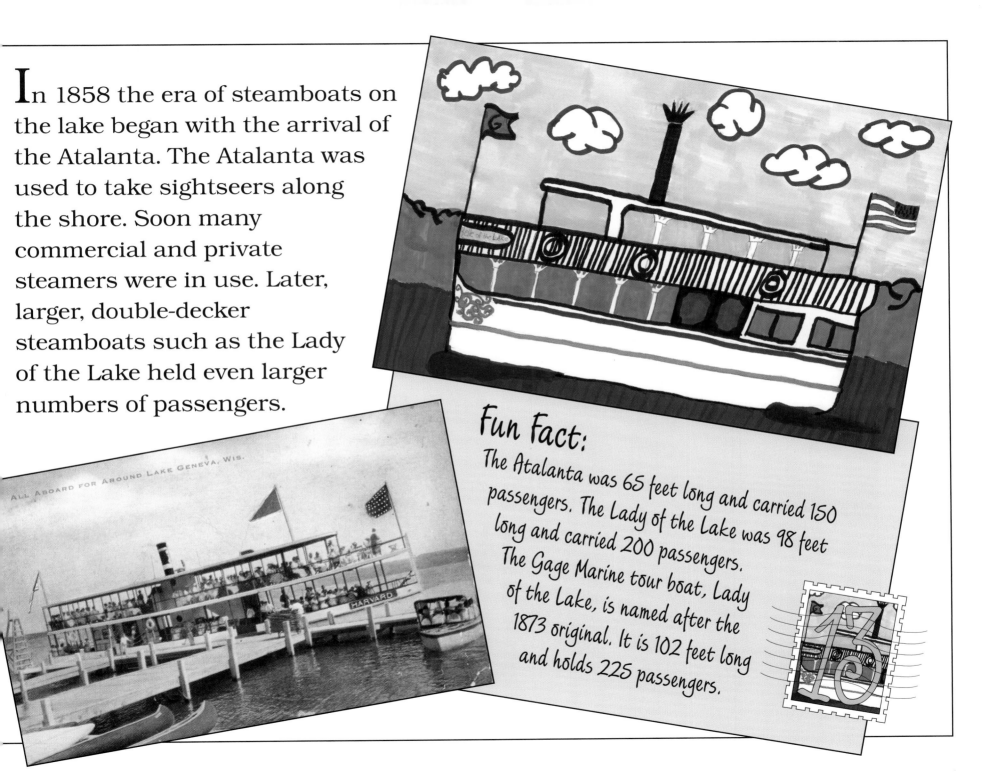

ALL ABOARD FOR AROUND LAKE GENEVA, WIS.

Fun Fact:

The Atalanta was 65 feet long and carried 150 passengers. The Lady of the Lake was 98 feet long and carried 200 passengers. The Gage Marine tour boat, Lady of the Lake, is named after the 1873 original. It is 102 feet long and holds 225 passengers.

In 1871, two things happened that would dramatically change life on Geneva Lake. In July, the first train on the newly built Chicago & Northwestern Railway arrived in Lake Geneva. Soon many more trains were bringing people to the lake.

Fun Fact:
By 1879, there were as many as 10 trains per day making this trip!

Lake Geneva, Wis. Arrival of Train at N. W. Ry. Station, William's Bay.

14

In October of 1871, the great Chicago fire burned down most of the city. Some of the people who lost their homes came to the lake to stay with friends and relatives while their homes were being rebuilt. Others decided to build new homes near or on the lake.

Many large estates were built on the lake during the late 1800's. You might recognize the names of some of the owners of these mansions, such as William Wrigley, who owned a company that made chewing gum, and Ignaz Schwinn, who owned a company that made bicycles.

Fun Fact:

The mansion shown here is Stone Manor. Stone Manor was built in 1899 by Otto Young, and is still standing today. At the time it was built, Stone Manor was the largest home on the lake, with 50 rooms on four floors!

Residence of Mrs. Otto Young, Lake Geneva, Wis.

The owners of these mansions would pick a summer evening to decorate their boats with candle-lit lanterns and parade their yachts around the lake. This evening came to be called "Venetian" night.

~Post Card~

Fun Fact:
Today the Lake Geneva Jaycees sponsor an annual Venetian Festival that ends with a lighted boat parade and fireworks.

Many hotels and resorts were also built along the lake. One resort, Kaye's Park was set on 300 acres and had its own farm, zoo and dance hall!

Fun Fact:
In 1895, 450 carloads of ice were shipped by train to Chicago.

While the lake served as a wonderful vacation spot for summer visitors, it also served as a refrigerator! In 1874, the first large ice house was built and men worked from January until early March cutting and hauling large blocks of ice. This ice was packed in the ice house between layers of sawdust, and then shipped by train to Chicago during the summer months.

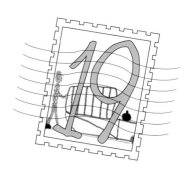

In 1897, a Chicago streetcar owner, Charles Yerkes, gave the money for the University of Chicago to build an observatory on a high point in Williams Bay. Astronomers used telescopes to take pictures of stars, galaxies, and planets.

Fun Fact:
Yerkes Observatory has a 40-inch refracting telescope, the largest in the world! It is 62 feet long and weighs 20 tons.

In 1916, a special boat was launched to carry the mail to people with homes on the lake. This mailboat, the Walworth, carried mail to lake shore residents for fifty years.

"Walworth"
Famous U. S. MAIL BOAT
Wisconsin Transportation Co.
Lake Geneva, Wisconsin

WALWORTH U.S.MAIL

Mail

Postmark

Fun Fact:
Today the Walworth II carries the U.S. mail during the summer months to about 50 residents who have their mail boxes at the lake shore.

21

As more people used the lake for swimming and boating, accidents occurred. In 1920, Simeon B. Chapin established the Water Safety Patrol. The Water Safety Patrol trained lifeguards and gave swimming lessons. Later, the Patrol bought boats that could rescue people in trouble on the lake.

Fun Fact:

Today the Water Safety Patrol is on call 24 hours a day during the summer months to help in any emergency that occurs on Geneva Lake.

The Water Safety Patrol picked up the passengers on the tour boat Majestic when it caught on fire on July 26, 1947. The next day the burned-out remains of the boat's hull was sunk in the deepest part of the lake.

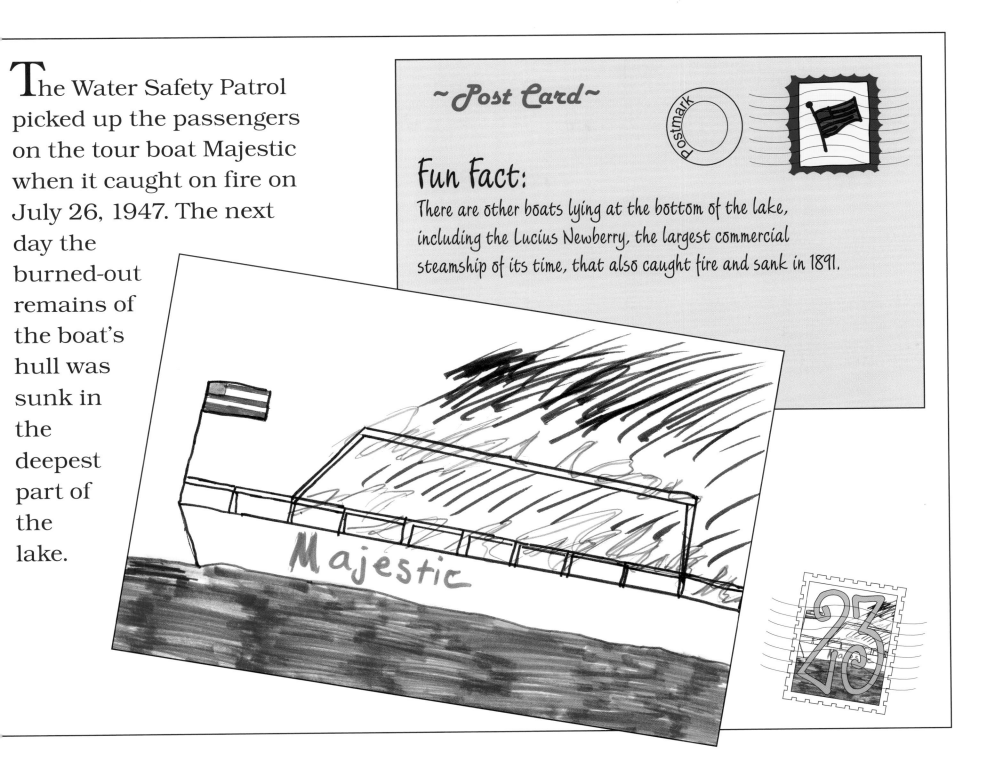

~Post Card~

Postmark

Fun Fact:

There are other boats lying at the bottom of the lake, including the Lucius Newberry, the largest commercial steamship of its time, that also caught fire and sank in 1891.

Majestic

23

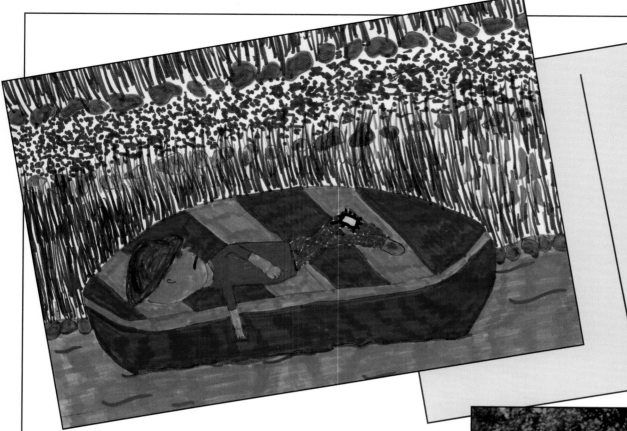

Fun Fact:
You can still walk along the entire shore path, even where it crosses private property, but you must enter on public land.

Today thousands of people from all over the world visit Geneva Lake. Some things are still as they were for the Potawatomis. You still see canoes on the lake. People still go fishing and walk along the shore path.

You won't see the huge prairies or as much of the wetlands as the Potawatomis saw. We've turned almost all of those areas into farms, towns, or roads. We've taken away part of the lake's renewal system and the habitat of many plants and animals.

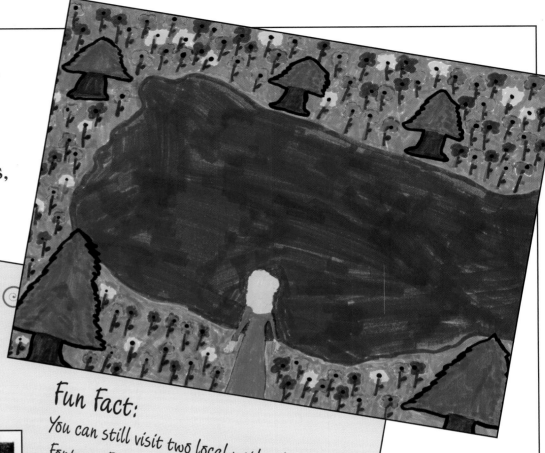

~Post Ca

1744 LAKE SHORE DRIVE NEAR PARK AND PICNIC GROUNDS, LAKE GENEVA, WISCONSIN

SA-H2670

Fun Fact:

You can still visit two local wetland areas. The Fontana Fen covers 10½ acres and contains the rarest type of wetland community in the world. The Kishwauketoe Conservancy in Williams Bay covers 231 acres and contains wetlands, dry prairies, and woodlands. Some shoreland areas are also being replanted to the beautiful prairie flowers and grasses.

25

Rural Scene, Lake Geneva, Wis.

While we enjoy Geneva Lake, we need to remember that our presence here is only one chapter in a much longer story. Just like the Potawatomis, we share the lake with many other living things.

It's up to all of us to make sure that the story of Geneva Lake continues and that it has a happy ending for future generations.

Geneva Lake Conservancy

Your Link to the Future

Postcards from the Lake tells the story of one of the most recognized bodies of water in southeastern Wisconsin. The very qualities that make living near or simply visiting Geneva Lake so desirable, however, are threatened by the rapid pace at which area development is occurring. As others have said, "You can love a good thing to death."

The Geneva Lake Conservancy exists to make sure that does not happen. Working with private landowners, government officials, developers and others, the Conservancy is a leading voice in Walworth County for the proper zoning, use, management and protection of private and public lands. Responsible stewardship – balancing the need for protection of our lakes, prime farmland, environmental corridors and other natural areas, with the demands for residential and commercial development - is the mission we have undertaken on behalf of Walworth County residents, and visitors too. We want your children and grandchildren, and their grandchildren, to send and receive future postcards from the lake that reflect our job well done.

The Conservancy, a not-for-profit charitable organization, is supported by donations from persons who share our vision for this special place. To learn more about how you can help conserve the lands and waters of beautiful Walworth County, please call us at (262) 275-5700. Or discover your link to the future in another way by logging on to our Internet web site at **www.genevalakeconservancy.org**

Douglass-Stevenson Fontana Mill House – Built 1857

Help Conserve the Lands and Waters of Beautiful Walworth County

P.O. Box 588 • 398 Mill Street • Fontana, WI 53125
Phone (262) 275-5700 • Fax (262) 275-0579 • Email GLC@genevaonline.com

Geneva Lake

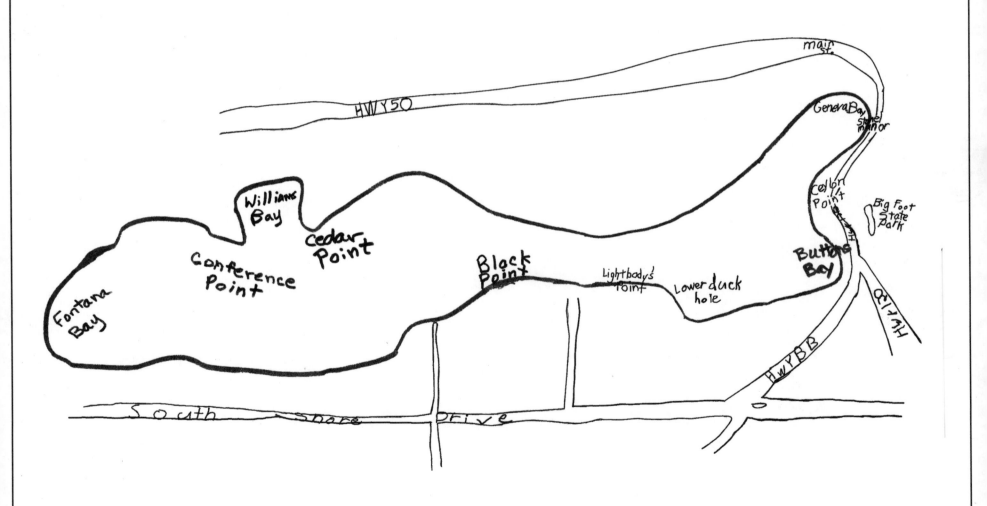

Williams Bay

Cedar Point

Conference Point

Fontana Bay

HWY 50

main St.

Geneva Bay

Stone manor

Ceylon Point

Big Foot State Park

Buttons Bay

Block Point

Lightbody's Point

Lower duck hole

HWY 120

HWY BB

South Shore Drive